Joseph Priestley

The discovery of oxygen

Part 1. : experiments

Joseph Priestley

The discovery of oxygen
Part 1. : experiments

ISBN/EAN: 9783742827173

Manufactured in Europe, USA, Canada, Australia, Japa

Cover: Foto ©berggeist007 / pixelio.de

Manufactured and distributed by brebook publishing software
(www.brebook.com)

Joseph Priestley

The discovery of oxygen

THE
DISCOVERY OF OXYGEN,

PART I.

EXPERIMENTS BY

JOSEPH PRIESTLEY, LL.D.

(1775.)

———————

EDINBURGH:

WILLIAM F. CLAY, 18 TEVIOT PLACE.

LONDON:

SIMPKIN, MARSHALL, HAMILTON, KENT & CO. LTD.

1894.

PREFACE.

THE extraordinary industry and productiveness of PRIESTLEY in the Department of "Pneumatic Chemistry," formed one of the most remarkable features of the period of Chemical investigation immediately preceding and contemporaneous with the establishment of the antiphlogistic system by Lavoisier, and served to promote the more particular examination of the properties of the gases by various workers. PRIESTLEY's discovery of oxygen is unquestionably the most important of all his numerous discoveries in this department, and its being made at the time it was, furnished Lavoisier with an opportune weapon with which to wage war against the Phlogiston theory.

It is impossible to overlook the fact that many of Priestley's experimental results were highly inaccurate, and that his conclusions, often too hastily drawn, were frequently erroneous ; but at the same time his ingenuity in devising apparatus, and his skill in carrying out experiments, were very remarkable, and placed him amongst the foremost philosophers of the day.

In the three sections of the second volume of his Observations on Air reprinted here, Priestley gave the first published account of the discovery of oxygen.

L. D.

ON DEPHLOGISTICATED AIR.*

SECTION III.

OF DEPHLOGISTICATED AIR, AND OF THE CONSTITUTION OF THE ATMOSPHERE.

THE contents of this section will furnish a very striking illustration of the truth of a remark, which I have more than once made in my philosophical writings, and which can hardly be too often repeated, as it tends greatly to encourage philosophical investigations viz. that more is owing to what we call *chance*, that is, philosophically speaking, to the observation of *events arising from unknown causes*, than to any proper *design*, or pre-conceived *theory* in this business. This does not appear in the works of those who write *synthetically* upon these subjects; but would, I doubt not, appear very strikingly in those who are the most celebrated for their philosophical acumen, did they write *analytically* and ingenuously.

For my own part, I will frankly acknowledge, that, at the commencement of the experiments recited in this section, I was so far from having formed any hypothesis that led to the discoveries I made in pursuing them, that they would have appeared very improbable to me had I been told of them; and when the decisive facts did at

---

* From Experiments and Observations on Different Kinds of Air. Vol. II. By Joseph Priestley, LL.D., F.R.S. London: 1775. Sections III.-V.; pp. 29-103.

length obtrude themselves upon my notice, it was very slowly, and with great hesitation, that I yielded to the evidence of my senses. And yet, when I re-consider the matter, and compare my last discoveries relating to the constitution of the atmosphere with the first, I see the closest and the easiest connexion in the world between them, so as to wonder that I should not have been led immediately from the one to the other. That this was not the case, I attribute to the force of prejudice, which, unknown to ourselves, biasses not only our *judgments*, properly so called, but even the perceptions of our senses: for we may take a maxim so strongly for granted, that the plainest evidence of sense will not intirely change, and often hardly modify our persuasions; and the more ingenious a man is, the more effectually he is entangled n his errors; his ingenuity only helping him to deceive himself, by evading the force of truth.

There are, I believe, very few maxims in philosophy that have laid firmer hold upon the mind, than that air, meaning atmospherical air (free from various foreign matters, which were always supposed to be dissolved, and intermixed with it) is *a simple elementary substance*, indestructible, and unalterable, at least as much so as water is supposed to be. In the course of my inquiries, I was, however, soon satisfied that atmospherical air is not an unalterable thing; for that the phlogiston with which it becomes loaded from bodies burning in it, and animals breathing it, and various other chemical processes, so far alters and depraves it, as to render it altogether unfit for inflammation, respiration, and other purposes to which it is subservient; and I had discovered that agitation in water, the process of vegetation, and probably other natural processes, by taking out the superfluous phlogiston, restore it to its original purity. But I own I had no idea of the possibility of going any farther in this way, and thereby procuring air purer than the best common air. I might, indeed, have naturally imagined that such would be air that should contain less phlogiston than the air of the atmosphere; but I had no idea that such a composition was possible.

It will be seen in my last publication, that, from the experiments which I made on the marine acid air, I was led to conclude, that common air consisted of some acid (and I naturally inclined to the acid that I was then operating upon) and phlogiston; because the union of this acid vapour and phlogiston made inflammable air; and inflammable air, by agitation in water, ceases to be inflammable, and becomes respirable. And though I could never make it quite so good as common air, I thought it very probable that vegetation, in more favourable circumstances than any in which I could apply it, or some other natural process, might render it more pure.

Upon this, which no person can say was an improbable supposition, was founded my conjecture, of volcanos having given birth to the atmosphere of this planet, supplying it with a permanent air, first inflammable, then deprived of its inflammability by agitation in water, and farther purified by vegetation.

Several of the known phenomena of the *nitrous acid* might have led me to think, that this was more proper for the constitution of the atmosphere than the marine acid: but my thoughts had got into a different train, and nothing but a series of observations, which I shall now distinctly relate, compelled me to adopt another hypothesis, and brought me, in a way of which I had then no idea, to the solution of the great problem, which my reader will perceive I have had in view ever since my discovery that the atmospherical air is alterable, and therefore that it is not an elementary substance, but a *composition* viz. what this composition is, or *what is the thing that we breathe*, and how is it to be made from its constituent principles.

At the time of my former publication, I was not possessed of a *burning lens* of any considerable force; and for want of one, I could not possibly make many of the experiments that I had projected, and which, in theory, appeared very promising. I had, indeed, a *mirror* of force sufficient for my purpose. But the nature of this instrument is such, that it cannot be applied, with effect, except upon substances that are capable of being suspended, or resting on a very slender support. It cannot be directed at all upon any substance in the form of *powder*, nor hardly upon any thing that requires to be put into a vessel of quicksilver; which appears to me to be the most accurate method of extracting air from a great variety of substances, as was explained in the Introduction to this volume. But having afterwards procured a. lens of twelve inches diameter, and twenty inches focal distance, I proceeded with great alacrity to examine, by the help of it, what kind of air a great variety of substances, natural and factitious, would yield, putting them into the vessels represented fig. *a*, which I filled with quicksilver, and kept inverted in a bason of the same. Mr. Warltire, a good chymist, and lecturer in natural philosophy, happening to be at that time in Calne, I explained my views to him, and was furnished by him with many substances, which I could not otherwise have procured.

With this apparatus, after a variety of other experiments, an account of which will be found in its proper place, on the 1st of August, 1774, I endeavoured to extract air from *mercurius calcinatus per se*; and I presently found that, by means of this lens, air was expelled from it very readily. Having got about three or four times as much as the bulk of my materials, I admitted

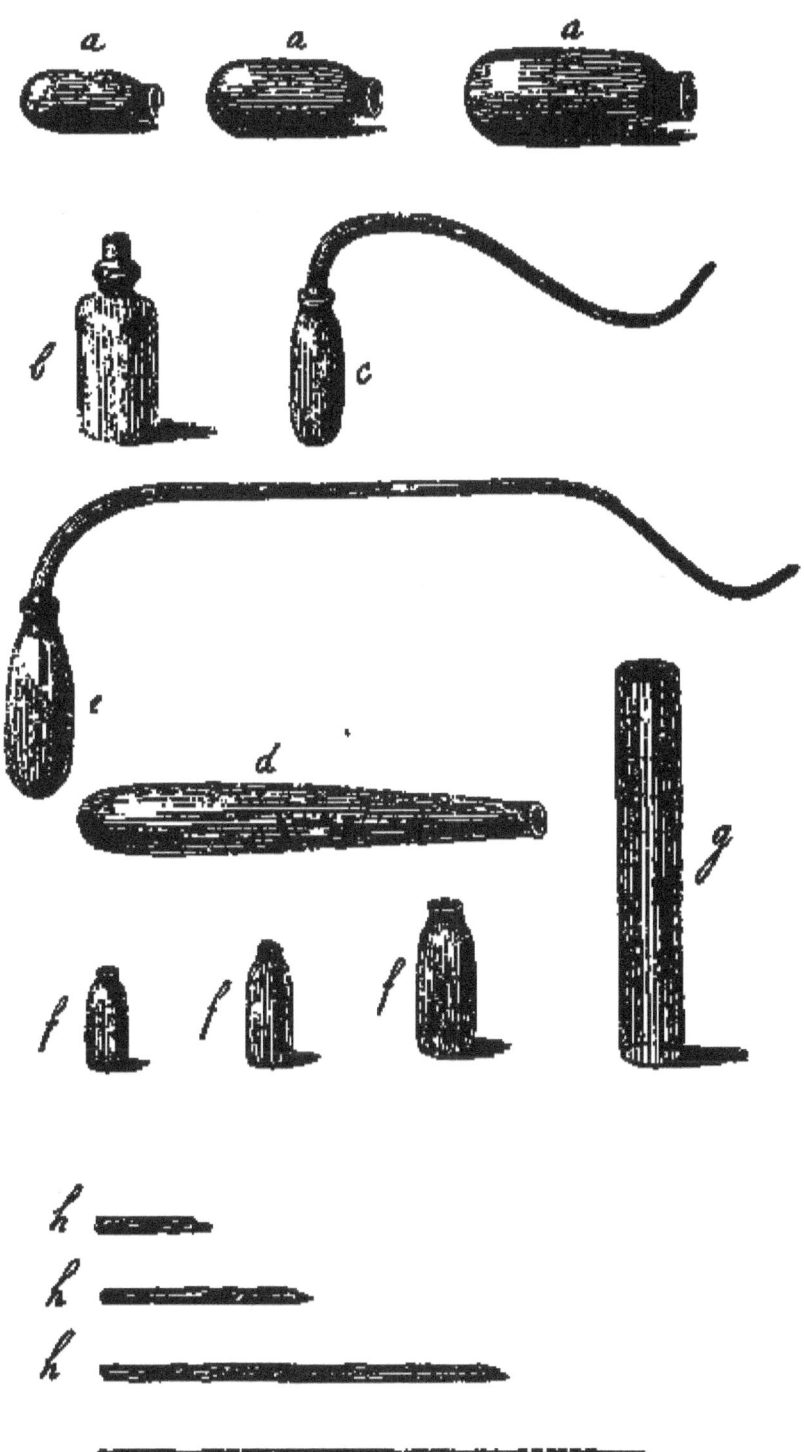

water to it, and found that it was not imbibed by it. But what surprized me more than I can well express, was, that a candle burned in this air with a remarkably vigorous flame, very much like that enlarged flame with which a candle burns in nitrous air, exposed to iron or liver of sulphur; but as I had got nothing like this remarkable appearance from any kind of air besides this particular modification of nitrous air, and I knew no nitrous acid was used in the preparation of *mercurius calcinatus*, I was utterly at a loss how to account for it.

In this case, also, though I did not give sufficient attention to the circumstance at that time, the flame of the candle, besides being larger, burned with more splendor and heat than in that species of nitrous air; and a piece of red-hot wood sparkled in it, exactly like paper dipped in a solution of nitre, and it consumed very fast; an experiment which I had never thought of trying with nitrous air.

At the same time that I made the above mentioned experiment, I extracted a quantity of air, with the very same property, from the common *red precipitate*, which being produced by a solution of mercury in spirit of nitre, made me conclude that this peculiar property, being similar to that of the modification of nitrous air above mentioned, depended upon something being communicated to it by the nitrous acid; and since the *mercurius calcinatus* is produced by exposing mercury to a certain degree of heat, where common air has access to it, I likewise concluded that this substance had collected something of *nitre*, in that state of heat, from the atmosphere.

This, however, appearing to me much more extraordinary than it ought to have done, I entertained some suspicion that the mercurius calcinatus, on which I had made my experiments, being bought at a common apothecary's, might, in fact, be nothing more than red precipitate; though, had I been anything of a practical chymist, I could not have entertained any such suspicion. However, mentioning this suspicion to Mr. Warltire, he furnished me with some that he had kept for a specimen of the preparation, and which, he told me, he could warrant to be genuine. This being treated in the same manner as the former, only by a longer continuance of heat, I extracted much more air from it than from the other.

This experiment might have satisfied any moderate sceptic : but, however, being at Paris in the October following, and knowing that there were several very eminent chymists in that place, I did not omit the opportunity, by means of my friend Mr. Magellan, to get an ounce of mercurius calcinatus prepared by Mr. Cadet, of the genuineness of which there could not possibly be any suspicion; and at the same time, I frequently mentioned my surprize at the kind of air which I had got from this preparation to Mr. Lavoisier, Mr. le Roy, and several other philosophers, who honoured me with their notice in that city; and who, I daresay, cannot fail to recollect the circumstance.

At the same time, I had no suspicion that the air which I had got from the mercurius calcinatus was even wholesome, so far was I from knowing what it was that I had really found; taking it for granted, that it was nothing more than such kind of air as I had brought nitrous air to be by the processes above mentioned; and in this air I have observed that a candle would burn sometimes quite naturally, and sometimes with a beautiful enlarged flame, and yet remain perfectly noxious.

At the same time that I had got the air above mentioned from mercurius calcinatus and the red precipitate, I had got the same kind from *red lead* or *minium*. In this process, that part of the minium on which the focus of the lens had fallen, turned yellow. One third of the air, in this experiment, was readily absorbed by water, but, in the remainder, a candle burned very strongly, and with a crackling noise.

That fixed air is contained in red lead I had observed before; for I had expelled it by the heat of a candle, and had found it to be very pure. (Vol. I. p. 192.) I imagine it requires more heat than I then used to expel any of the other kind of air.

This experiment with *red lead* confirmed me more in my suspicion, that the *mercurius calcinatus* must get the property of yielding this kind of air from the atmosphere, the process by which that preparation, and this of red lead is made, being similar. As I never make the least secret of any thing that I observe, I mentioned this experiment also, as well as those with the mercurius calcinatus, and the red precipitate, to all my philosophical acquaintances at Paris, and elsewhere; having no idea, at that time, to what these remarkable facts would lead.

Presently after my return from abroad, I went to work upon the *mercurius calcinatus*, which I had procured from Mr. Cadet; and, with a very moderate degree of heat, I got from about one fourth of an ounce of it, an ounce-measure of air, which I observed to be not readily imbibed, either by the substance itself from which it had been expelled (for I suffered them to continue a long time together before I transferred the air to any other place) or by water, in which I suffered this air to stand a considerable time before I made any experiment upon it.

In this air, as I had expected, a candle burned with a vivid flame; but what I observed new at this time (Nov. 19), and which surprized me no less than the fact I had discovered before, was, that, whereas a few moments agitation in water will deprive the modified nitrous air of its property of admitting a candle to burn in it; yet, after more than ten times as much agitation as would be sufficient to produce this alteration in the nitrous air, no sensible change was produced in this. A candle still burned in it with a strong flame; and it did not, in the least, diminish common air, which I have observed that nitrous air, in this state, in some measure, does.

But I was much more surprized, when, after two days, in which this air had continued in contact with water (by which it was diminished about one twentieth of its bulk) I agitated it violently in water about five minutes, and found that a candle still burned in it as well as in common air. The same degree of agitation would have made phlogisticated nitrous air fit for respiration indeed, but it would certainly have extinguished a candle.

These facts fully convinced me, that there must be a very material difference between the constitution of the air from mercurius calcinatus, and that of phlogisticated nitrous air, notwithstanding their resemblance in some particulars. But though I did not doubt that the air from *mercurius calcinatus* was fit for respiration, after being agitated in water, as every kind of air without exception, on which I had tried the experiment, had been. I still did not suspect that it was respirable in the first instance; so far was I from having any idea of this air being, what it really was, much superior, in this respect, to the air of the atmosphere.

In this ignorance of the real nature of this kind of air, I continued from this time (November) to the 1st of March following; having, in the mean time, been intent upon my experiments on the vitriolic acid air above recited, and the various modifications of air produced by spirit of nitre, an account of which will follow. But in the course of this month, I not only ascertained the nature of this kind of air, though very gradually, but was led by it to the complete discovery of the constitution of the air we breathe.

Till this 1st of March, 1775, I had so little suspicion of the air from mercurius calcinatus, &c. being wholesome, that I had not even thought of applying to it the test of nitrous air; but thinking (as my reader must imagine I frequently must have done) on the candle burning in it after long agitation in water, it occurred to me at last to make the experiment; and putting one measure of nitrous air to two measures of this air, I found, not only that it was diminished, but that it was diminished quite as much as common air, and that the redness of the mixture was likewise equal to that of a similar mixture of nitrous and common air.

After this I had no doubt but that the air from mercurius calcinatus was fit for respiration, and that it had all the other properties of genuine common air. But I did not take notice of what I might have observed, if I had not been so fully possessed by the notion of there being no air better than common air, that the redness was really deeper, and the diminution something greater than common air would have admitted.

Moreover, this advance in the way of truth, in reality, threw me back into error, making me give up the hypothesis I had first formed, viz. that the mercurius calcinatus had extracted spirit of nitre from the air; for I now concluded, that all the constituent parts of the air were equally, and in their proper proportion, imbibed in the preparation of this substance, and also in the process of making red lead. For at the same time that I made the above mentioned experiment on the air from mercurius calcinatus, I likewise observed that the air which I had extracted from red lead, after the fixed air was washed out of it, was of the same nature, being diminished by nitrous air like common air: but, at the same time, I was puzzled to find that air from the red precipitate was diminished in the same manner, though the process for making this substance is quite different from that of making the two others. But to this circumstance I happened not to give much attention.

I wish my reader be not quite tired with the frequent repetition of the word *surprize*, and others of similar import ; but I must go on in that style a little longer. For the next day I was more surprized than ever I had been before, with finding that, after the above-mentioned mixture of nitrous air and the air from mercurius calcinatus, had stood all night, (in which time the whole diminution must have taken place ; and, consequently, had it been common air, it must have been made perfectly noxious, and intirely unfit for respiration or inflammation) a candle burned in it, and even better than in common air.

I cannot, at this distance of time, recollect what it was that I had in view in making this experiment ; but I know I had no expectation of the real issue of it. Having acquired a considerable degree of readiness in making experiments of this kind, a very slight and evanescent motive would be sufficient to induce me to do it. If, however, I had not happened, for some other purpose, to have had a lighted candle before me, I should probably never have made the trial ; and the whole train of my future experiments relating to this kind of air might have been prevented.

Still, however, having no conception of the real cause of this phenomenon, I considered it as something very extraordinary ; but as a property that was peculiar to air that was extracted from these substances, and *adventitious ;* and I always spoke of the air to my acquaintance as being substantially the same thing with common air.

I particularly remember my telling Dr. Price, that I was myself perfectly satisfied of its being common air, as it appeared to be so by the test of nitrous air; though, for the satisfaction of others, I wanted a mouse to make the proof quite complete.

On the 8th of this month I procured a mouse, and put it into a glass vessel, containing two ounce-measures of the air from mercurius calcinatus. Had it been common air, a full-grown mouse, as this was, would have lived in it about a quarter of an hour. In this air, however, my mouse lived a full half hour; and though it was taken out seemingly dead, it appeared to have been only exceedingly chilled; for, upon being held to the fire, it presently revived, and appeared not to have received any harm from the experiment.

By this I was confirmed in my conclusion, that the air extracted from mercurius calcinatus, &c. was, *at least*, *as good* as common air; but I did not certainly conclude that it was any *better;* because, though one mouse would live only a quarter of an hour in a given quantity of air, I knew it was not impossible but that another mouse might have lived in it half an hour; so little accuracy is there in this method of ascertaining the goodness of air: and indeed I have never had recourse to it for my own satisfaction, since the discovery of that most ready, accurate, and elegant test that nitrous air furnishes. But in this case I had a view to publishing the most generally-satisfactory account of my experiments that the nature of the thing would admit of.

This experiment with the mouse, when I had reflected upon it some time, gave me so much suspicion that the air into which I had put it was better than common air, that I was induced, the day after, to apply the test of nitrous air to a small part of that very quantity of air which the mouse had breathed so long; so that, had it been common air, I was satisfied it must have been very nearly, if not altogether, as noxious as possible, so as not to be affected by nitrous air; when, to my surprize again, I found that though it had been breathed so long, it was still better than common air. For after mixing it with nitrous air, in the usual proportion of two to one, it was diminished in the proportion of $4\frac{1}{2}$ to $3\frac{1}{2}$; that is, the nitrous air had made it two ninths less than before, and this in a very short space of time; whereas I had never found that, in the longest time, any common air was reduced more than one fifth of its bulk by any proportion of nitrous air, nor more than one fourth by any phlogistic process whatever. Thinking of this extraordinary fact upon my pillow, the next morning I put another measure of nitrous air to the same mixture, and, to my utter astonishment, found that it was farther diminished to almost one half of its original quantity. I then put a third measure to it; but this did not diminish it any farther: but, however, left it one measure less than it was even after the mouse had been taken out of it.

Being now fully satisfied that this air, even after the mouse had breathed it half an hour, was much better than common air; and having a quantity of it still left, sufficient for the experiment, viz. an ounce-measure and a half, I put the mouse into it; when I observed that it seemed to feel no shock upon being put into it, evident signs of which would have been visible, if the air had not been very wholesome; but that it remained perfectly at its ease another full half hour, when I took it out quite lively and vigorous. Measuring the air the next day, I found it to be reduced from $1\frac{1}{2}$ to $\frac{2}{3}$ of an ounce-measure. And after this, if I remember well (for in my *register* of the day I only find it noted, that it was *considerably diminished* by nitrous air) it was nearly as good as common air. It was evident, indeed, from the mouse having been taken out quite vigorous, that the air could not have been rendered very noxious.

For my farther satisfaction I procured another mouse, and putting it into less than two ounce-measures of air extracted from mercurius calcinatus and air from red precipitate (which, having found them to be of the same quality, I had mixed together) it lived three quarters of an hour. But not having had the precaution to set the vessel in a warm place, I suspect that the mouse died of cold. However, as it had lived three times as long as it could probably have lived in the same quantity of common air, and I did not expect much accuracy from this kind of test, I did not think it necessary to make any more experiments with mice.

Being now fully satisfied of the superior goodness of this kind of air, I proceeded to measure that degree of purity, with as much accuracy as I could, by the test of nitrous air; and I began with putting one measure of nitrous air to two measures of this air, as if I had been examining common air; and now I observed that the diminution was evidently greater than common air would have suffered by the same treatment. A second measure of nitrous air reduced it to two thirds of its original quantity, and a third measure to one half. Suspecting that the diminution could not proceed much farther, I then added only half a measure of nitrous air, by which it was diminished still more; but not much, and another half measure made it more than half of its original quantity; so that, in this case, two measures of this air took more than two measures of nitrous air, and yet remained less than half of what it was. Five measures brought it pretty exactly to its original dimensions.

At the same time, air from the *red precipitate* was diminished in the same proportion as that from *mercurius calcinatus,* five measures of nitrous air being received by two measures of this without any increase of dimensions. Now as common air takes about one half of its bulk of nitrous air, before it begins to receive any addition to its dimensions from more nitrous air, and this air took more than four half-measures before it ceased to be diminished by more nitrous air, and even five half-measures made no addition to its original dimensions, I conclude that it was between four and five times as good as common air. It will be seen that I have since procured air better than this, even between five and six times as good as the best common air that I have ever met with.

Being now fully satisfied with respect to the *nature* of this new species of air, viz. that, being capable of taking more phlogiston from nitrous air, it therefore originally contains less of this principle ; my next inquiry was, by what means it comes to be so pure, or philosophically speaking, to be so much *dephlogisticated* ; and since the red lead yields the same kind of air with mercurius calcinatus, though mixed with fixed air, and is a much cheaper material, I proceeded to examine all the preparations of lead, made by heat in the open air, to see what kind of air they would yield, beginning with the *grey calx,* and ending with *litharge.*

The red lead which I used for this purpose yielded a considerable quantity of dephlogisticated air, and very little fixed air ; but to what circumstance in the preparation of this lead, or in the keeping of it, this difference is owing, I cannot tell. I have frequently found a very remarkable difference between different specimens of red lead in this respect, as well as in the purity of the air which they contain. This difference, however, may arise in a great measure, from the care that is taken to extract the fixed air from it. In this experiment two measures of nitrous air being put to one measure of this air, reduced it to one third of what it was at first, and nearly three times its bulk of nitrous air made very little addition to its original dimensions ; so that this air was exceedingly pure, and better than any that I had procured before.

The preparation called *massicot* (which is said to be a state between the grey calx and the red lead) also yielded a considerable quantity of air, of which about one half was fixed air, and the remainder was such, that when an equal quantity of nitrous air was put to it, it was something less than at first ; so that this air was about twice as pure as common air.

I thought it something remarkable, that in the preparations of lead by heat, those before and after these two, viz. the red lead and *massicot*, yielded only fixed air. I would also observe, by the way, that a very small quantity of air was extracted from *lead ore* by the burning lens. The bulk of it was easily absorbed by water. The remainder was not affected by nitrous air, and it extinguished a candle.

I got a very little air by the same process from the *grey calx of lead*, of precisely the same quality with the former. That part of it which was not affected by nitrous air extinguished a candle, so that both of them may be said to have yielded fixed air, only with a larger portion than usual, of that part of it which does not unite with water.

Litharge (which is a state that succeeds the red lead) yielded air pretty readily ; but this also was fixed air. That which was not absorbed by water, was not affected by nitrous air.

Much more than I had any opportunity of doing remains to be done, in order to ascertain upon what circumstances, in these preparations of lead, the quality of the air which they contain, depends. It can only be done by some person who shall carefully attend to the processes, so as to see himself in what manner they are made, and examine them in all their different states. I very much wished to have attempted something of this kind myself, but I found it impossible in my situation. However, I got Dr. Higgins (who furnished me with several preparations that I could not easily have procured elsewhere) to make me a quantity of red lead, that I might, at least, try it when *fresh made*, and after keeping it some time in different circumstances ; and though, by the help of this preparation, I did not do the thing that I expected, I did something else, much more considerable.

This fresh made red lead had a yellowish cast, and had in it several pieces intirely yellow. I tried it immediately, in the same manner in which I had made the preceding experiments, viz. with the burning lens in quicksilver, and found that it yielded very little air, and with great difficulty ; requiring the application of a very intense heat. With an equal quantity of nitrous air, a part of this air was reduced to one half of its original bulk, and $3\frac{1}{2}$ measures saturated it. The air, therefore, was very pure, and the quantity that it yielded being very small, it proved to be in a very favourable state for ascertaining on what circumstances its acquiring this air depended.

My object now was to bring this fresh made red lead, which yielded very little air, to that state in which other red lead had yielded a considerable quantity ; and taking it, in a manner, for granted, in consequence of the reasoning intimated above, that red lead must imbibe from the atmosphere some kind of acid, in order to acquire that property, I took three separate half-ounces of this fresh made red lead, and moistened them till they made a kind of paste, with each of the three mineral acids, viz. the vitriolic, the marine, and the nitrous ; and as I intended to make the experiments in a gun-barrel, lest the iron should be too much affected by them, I dried all these mixtures, till they were perfectly hard ; then pulverizing them, I put them separately into my gun-barrel, filled up to the mouth with pounded flint, which I had found by trial to yield little, or no air when treated in this ·manner. I had also found that no quantity of air, sufficient to make an experiment, could be procured from an equal quantity of this red lead by this process.

Those portions of the red lead which had been moistened with the vitriolic and marine acids became white ; out that which had been moistened with the nitrous acid, had acquired a deep brown colour. The mixtures with the nitrous and marine acids dried pretty readily, but that with the vitriolic acid was never perfectly dry ; but a great part of it remained in the form of a softish paste.

Neither the vitriolic nor the marine acid mixtures gave the least air when treated in the manner above mentioned ; but the moment that the composition into which the *nitrous* acid had entered became warm, air began to be produced ; and I received the produce in quicksilver. About one ounce-measure was quite transparent, but presently after it became exceedingly red ; and being satisfied that this redness was owing to the nitrous acid vapour having dissolved the quicksilver, I took no more than two ounce-measures in this way, but received all the remainder, which was almost two pints, in water. Far the greatest part of this was fixed air, being readily absorbed by water, and extinguishing a candle. There was, however, a considerable residuum, in which the flame of a candle burned with a crackling noise, from which I concluded that it was true dephlogisticated air.

In this experiment I had moistened the red lead with spirit of nitre several times, and had dried it again. When I repeated the experiment, I moistened it only once with the same acid, when I got from it not quite a pint of air; but it was almost all of the dephlogisticated kind, about five times as pure as common air. N.B.—All the acids made a violent effervescence with the red lead.

Though there was a difference in the result of these experiments, which I shall consider hereafter, I was now convinced that it was the nitrous acid which the red lead had acquired from the air, and which had enabled it to yield the dephlogisticated air, agreeable to my original conjecture. Finding also, as will be seen in the following section, that the same kind of air is produced by moistening with the spirit of nitre any kind of earth that is free from phlogiston, and treating it as I had done the red lead in the last-mentioned experiment, there remained no doubt in my mind, but that *atmospherical air,* or the thing that we breathe, *consists of the nitrous acid and earth,* with so much phlogiston as is necessary to its elasticity; and likewise so much more as is required to bring it from its state of perfect purity to the mean condition in which we find it.

For this purpose I tried, with success, *flowers of zinc, chalk, quick-lime, slacked-lime, tobacco-pipe clay, flint* and *Muscovy talck,* with other similar substances, which will be found to comprize all the kinds of earth that are essentially distinct from each other, according to their chymical properties. A particular account of the processes with these substances, I reserve for another section; thinking it sufficient in this to give a history of the discovery, and a general account of the nature of this dephlogisticated air, with this general inference from the experiment, respecting the constitution of the atmosphere.

I was the more confirmed in my idea of spirit of nitre and earth constituting respirable air, by finding, that when any of these matters, on which I had tried the experiment, had been treated in the manner above mentioned, and they had thereby yielded all the air that could be extracted from them by this process; yet when they had been moistened with fresh spirit of nitre, and were treated in the same manner as before, they would yield as much dephlogisticated air as at the first. This may be repeated till all the earthy matter be exhausted. It will be sufficient to recite one or two facts of this kind from my register.

April 18, I took the remains of the fresh made red lead, out of which a great quantity of dephlogisticated air had been extracted, and moistening about three quarters of an ounce of it a second time with spirit of nitre, I got from it about two pints of air, all of which was nearly six times as pure as common air. This air was generated very fast, and the glass tube through which it was transmitted was filled with red fumes; the nitrous acid, I suppose, prevailing in the composition of the air, but being absorbed by the water in which it was afterwards received.

In this, and many other processes, my reader will find a great variety in the purity of the air procured from the same substances. But this will not be wondered at, if it be considered that a small quantity of phlogistic matter, accidentally mixing with the ingredients for the composition of this air, depraves it. It will also be unavoidably depraved, in some measure, if the experiment be made in a gun-barrel, which I commonly made use of, when, as was generally the case, it was sufficiently exact for my purpose, on account of its being the easiest, and in many respects, the most commodious process.

The reason of this is, that if the produce of air be not

very rapid, there will be time for the phlogiston to be disengaged from the iron itself, and to mix with the air. Accordingly I have seldom failed to find, that when I endeavoured to get all the air I possibly could from any quantity of materials, and received the produce at different times (as for my satisfaction I generally did) the last was inferior in purity to that which came first. Not unfrequently it was phlogisticated air; that is, air so charged with phlogiston, as to be perfectly noxious; and sometimes, as the reader will find in the next section, it was even nitrous air.

On the same account it frequently happened, that when I used a considerable degree of heat, the red lead which I used in these experiments would be changed into real lead, from which it was often very difficult to get the gun-barrel perfectly clear.

A good deal will also depend upon the ingredients which have been used in the gun-barrel in preceding experiments : for it is not easy to get such an instrument perfectly clean from all the matters that have been put into it ; and though it may be presumed, in general, that every kind of air will be expelled from such ingredients by making the tube red-hot; yet matters containing much phlogiston, as charcoal, &c. will not part with it in consequence of the application of heat, unless there be at hand some other substance with which it may combine. Though, therefore, a gun-barrel, containing such small pieces of charcoal as cannot be easily wiped out of it, be kept a long time in a red heat, and even with its mouth open ; yet if it be of a considerable length, some part of the charcoal may remain unconsumed, and the effect of it will be found in the subsequent experiment. Of this I had the following very satisfactory proof.

Being desirous to shew some of my friends the actual production of dephlogisticated air, and having no other apparatus at hand, I had recourse to my gun-barrel; but apprized them, that having used it the day before to get air from charcoal, with which it had been filled for that purpose, though I had taken all the pains I could to get it all out, yet so much would probably remain, that I could not depend on the air I should get from it being dephlogisticated; but that it would probably be of an inferior quality, and perhaps even nitrous air. Accordingly, having put into it a mixture of spirit of nitre and red lead (being part of a quantity which I had often used before for the same purpose) dried, and pounded, I put it into the fire, and received the air in water.

The first produce, which was about a pint, was so far nitrous, that two measures of common air, and one of this, occupied the space of little more than two measures; that is, it was almost as strongly nitrous as that which is produced by the solution of metals in spirit of nitre. The second pint was very little different from common air, and the last produce was better still, being more than twice as good as common air. If, therefore, any person shall propose to make dephlogisticated air, in large quantities, he should have an apparatus appropriated to that purpose; and the greatest care should be taken to keep the instruments as clear as possible from all phlogistic matter, which is the very bane of purity with respect to air, they being exactly *plus* and *minus* to each other.

The hypothesis maintained in this section, viz. that atmospherical air consists of the nitrous acid and earth, suits exceedingly well with the facts relating to the production of nitre; for it is never generated but in the open air, and by exposing to it such kinds of earth as are known to have an affinity with the nitrous acid; so that by their union common nitre may be formed.

Hitherto it has been supposed by chymists, that this

nitrous acid, by which common nitre is formed, exists in the atmosphere as an *extraneous substance*, like water, and a variety of other substances, which float in it, in the form of effluvia; but since there is no place in which nitre may not be made, it may, I think, with more probability be supposed, according to my hypothesis, that nitre is formed by a real *decomposition of the air itself*, the *bases* that are presented to it having, in such circumstances, a nearer affinity with the spirit of nitre than that kind of earth with which it is united in the atmosphere.

My theory also supplies an easy solution of what has always been a great difficulty with chymists, with respect to the *detonation of nitre.* The question is, what becomes of the nitrous acid in this case? The general, I believe the universal, opinion now is, that it is *destroyed*; that is, that the acid is properly decomposed, and resolved into its original elements, which Stahl supposed to be earth and water. On the other hand, I suppose that, though the common properties of the acid, as combined with water, disappear, it is only in consequence of its combination with some earthy or inflammable matter, with which it forms some of the many species of air, into the composition of which this wonderful acid enters. It may be common air, it may be dephlogisticated air, or it may be nitrous air, or some of the other kinds, of which an account will be given in a subsequent section. That it should really be the nitrous acid, though so much disguised by its union with earthy, or other matters, will not appear extraordinary to any person who shall consider how little the acid of vitriol is apparent in common sulphur.

With respect to *mercurius calcinatus*, and *red lead*, their red colour favours the supposition of their having extracted spirit of nitre from the air.

SECTION IV.

A MORE PARTICULAR ACCOUNT OF SOME PROCESSES FOR THE PRODUCTION OF DEPHLOGISTICATED AIR.

I CANNOT promise those of my readers, whose object is nothing more than *general information*, much pleasure from the perusal of this section, as it will consist, for the most part, of a dry detail of processes, for procuring dephlogisticated air; but as they all appeared necessary, in my investigation of the subject, I doubt not but an attention to them will be of use to such as are disposed to pursue the inquiry themselves. I might have contented myself with giving a general idea of the result of such experiments; but that would have been to mix my own *opinions* with *facts*, in such a manner that the reader would not have been able to separate them. At present, if I should be mistaken in any of my opinions, the reader, having before him all the facts on which those opinions were grounded, will be able to rectify the mistake, and prevent the error from spreading.

Having seen sufficient reason to conclude that respirable air consists of nitrous acid and earth, my object, in all this course of experiments, was simply to find *what kind of earth* was most proper for this purpose, or which had the most aptness to form this peculiar union with the nitrous acid. Upon the whole, I think it will appear that the *metallic earths*, if they be free from phlogiston, are the most proper, and next to them the *calcareous* earths; but that a very great difference in the production of this kind of air depends upon a variety of circumstances in which the experiments are made.

I have observed that *red lead*, without any addition, yields dephlogisticated air by heat. To give some idea of the differences in the results, from what is, to appearance, the same preparation, and of the consequence of adding spirit of nitre to the red lead, I must inform my reader, that having weighed two half-ounces of red lead, taken from the same parcel, I put one of them, without any addition, into the gun-barrel, and with a very brisk fire (which is generally a considerable advantage for the production of air) I got no more than three ounce-measures, and it was very little better than common air.

The second half-ounce I moistened with a very diluted spirit of nitre; and when it was dried and pounded, I put it into the same gun-barrel; and, in the same circumstances with the former, I got from it about three pints of air, the first part of which was so far dephlogisticated, that two measures of it, and five of nitrous air, occupied the space of two measures only; of the second quantity, two measures were not increased by the addition of seven measures of nitrous air. This was the purest air that I had then seen. The last produce was almost all pure fixed air, being not at all affected by nitrous air, extinguishing a candle, and precipitating lime in lime-water. It was, indeed, a little of a nitrous nature; for it diminished common air in a small degree, an effect which I attribute to the phlogiston coming from the iron.

A remarkable difference in the quantity of the produce of this kind of air, as I hinted just now (and as I have observed in a former publication, in the produce of inflammable air) depends upon the *suddenness* with which the same degree of heat is applied. The following must be reckoned a remarkable fact of this kind, and it was made with as much care as I could possibly apply. From an ounce of red lead, by a sudden and brisk heat, I got above two quarts of air, a great part of which was fixed air, and the rest was about twice as good as common air; and immediately after, putting the very same quantity of the same parcel of red lead into the same gun-barrel, by heat very *slowly applied*, but urged vehemently at last, I got no more than two ounce-measures of air, a great part of which was fixed air, and the rest not so good as common air.

I had been told that red lead acquired additional weight by being often washed in water. In order to try whether this was the fact, and also whether the red lead acquired its property of yielding dephlogisticated air by this means, I washed a quantity of that parcel which I had got fresh-made, four times in distilled water, evaporating it to dryness each time; but no more air came from it than when it had not been wetted, neither was it at all increased in weight by this means.

I have observed that, in general, those substances which, without containing phlogiston, yield fixed air with heat, or by the addition of an acid, when mixed with spirit of nitre, and treated as above, yield more or less of dephlogisticated air; but generally with a considerable mixture of fixed air, though I profess not to know upon what circumstance it is that the proportion of these two kinds of air, produced from these substances, depends.

With a very small degree of heat, *white lead*, without any addition, yields a very great quantity of pure fixed air. Having moistened about an ounce-measure of it with spirit of nitre, and having put it into a glass-vessel, with a ground-stopple and tube, I extracted from it, at five different times, five pints of air, each of which I examined separately, as usual, and the results were as follows.

Of the first quantity, about $\frac{19}{20}$ or $\frac{59}{60}$ was absorbed by water, and the remainder neither affected common air, nor was affected by nitrous air, so that it was pure fixed air; and considering the quantity that would be necessarily absorbed by the water in which it was received, before I made any trial of the properties of it, it may perhaps be deemed to have been as free from any foreign mixture as any that was ever procured. Of the second quantity, about twice as much was left unabsorbed by water, and the residuum appeared to be dephlogisticated; for it took about an equal measure of nitrous air to saturate it; and consequently it was nearly twice as good as common air. Of the third quantity, as little remained unabsorbed by water as of the first; but the residuum was as pure as that of the second quantity. Of the fourth quantity, one-fourth remained unabsorbed by water, and it took $1\frac{3}{4}$ of nitrous air to saturate it. Of the fifth pint, one-half remained unabsorbed, and it took more than two equal measures of nitrous air to saturate it; so that it was nearly four times as pure as common air. Lastly, a single ounce-measure, that came very slowly after the five pints, was no part of it absorbed by water, and it took $1\frac{1}{2}$ of nitrous air to saturate it, being about three times as pure as common air.

From a quantity of *litharge*, moistened with spirit of nitre, and dried, I got, in a gun-barrel, a great quantity of air; about half of which was fixed air, precipitating lime in lime-water, and the other half was strongly nitrous; but with a burning lens, in quicksilver, I got a very pure dephlogisticated air from this mixture.

To go through the different states of lead in this manner, I took half an ounce of *lead-ore*, and having saturated it with spirit of nitre, I dried it as before, put it into a gun-barrel, filled up to the mouth with pounded flint, and placed vessels filled with water to receive the air. The consequence was, that as soon as this mixture began to be warm, air was generated very fast, insomuch that, being rather alarmed, I stood on one side; when presently there was a violent and loud explosion, by which all the contents of the gun-barrel were driven out with great force, dashing to pieces the vessels that were placed to receive the air, and dispersing the fragments all over the room; so that all the air which I had collected, and which was about a pint, was lost. The mixture, before it was put into the gun-barrel, was betwixt white and yellow, and had very much the smell of brimstone; so that it was in fact a composition similar to gunpowder.

Being desirous to know what kind of air I had got by this process, I put the same materials into a glass-phial, and putting it into a crucible with sand, disposed the apparatus for receiving the air in such a manner, that the explosion could not affect it. It did explode as before, but the air was preserved, and appeared to be very strong nitrous air, almost as much so as that which is procured by the solution of metals.

From the *grey calx of lead,* treated in the same manner, I got about a pint of air, half of which, being readily absorbed by water, I take for granted, was fixed air; but the remainder was strongly nitrous. Had I not washed it a good deal in water, it would probably have been as strong as that which is procured from metals.

The purest air that I ever procured was from *flowers of zinc,* moistened, as in the other processes, with spirit of nitre, and put into a glass-phial, with a ground-stopple and tube. At first I despaired of getting any air at all from the process; but at length it came in a prodigious torrent, and was so cloudy, that the bursting of every bubble, after it had passed through the water, resembled the bursting of a bag of flour. The tube through which it was transmitted was exceedingly red, and in some degree, the inside of the receiver too, as might be perceived amidst the thick cloud that filled it. This cloudiness of the newly-generated air, I have often perceived in the process with red-lead, but never in so great a degree as in this case.

The quantity of air procured was nearly three pints, from about half an ounce-measure of the flowers of zinc; and it was so highly dephlogisticated, that it took three times its bulk of nitrous air before its dimensions were increased. When it had got only twice its bulk of nitrous air, it was reduced to less than one-fifth of its original quantity. The last produce came very slowly, and was not quite so pure. The flowers of zinc, which I used in this experiment, I had from Dr. Higgins. They formed a very hard and brittle substance, when mixed with spirit of nitre, and dried. After the process it swelled, and broke the phial into many pieces.

Beside these, I tried no earth of any metal except the *rust of iron and white arsenic*, both of which, when treated in the manner above mentioned, and put into a gun-barrel, yielded nothing but fixed or nitrous air; so that these calces undoubtedly contained much phlogiston, and the flowers of zinc, perhaps, none at all.

From considerably less than half an ounce of *rust of iron*, moistened with spirit of nitre, and dried, I got about a quart of air, about one-third of which was fixed air, precipitating lime in lime-water, &c. and the remainder was nitrous; so that two measures of common air and one of this, occupied the space of less than two measures.

The *white arsenic* I procured from Dr. Higgins, who assured me that it contained the least phlogiston possible. I moistened about an ounce-measure of it with spirit of nitre, and putting it into a phial with a ground-stopple and tube, with no great degree of heat, I extracted from it four ounce-measures of air, and it was as strongly nitrous as any that I had ever procured from metals. I increased the heat till the phial was melted, without getting any more air. The tube was exceedingly red during the transmission of the air through it.

Next to the metallic earths of lead and zinc, I found the *calcareous* earths the most proper for the production of dephlogisticated air; but I had no opportunity of trying any great variety of them. The best that I did try was *chalk.* Having saturated half an ounce of it with diluted spirit of nitre, and dried it, I got from it, in a gun-barrel, more than a pint of air, which was highly dephlogisticated. I began to receive this produce in quicksilver, the consequence of which was, that the nitrous acid, coming over in the form of vapour, dissolved the quicksilver, and made nitrous air; but a crust being formed upon the surface of it, prevented the solution of more, and the air continued red a long time.

From another ounce-measure of chalk, treated in the same manner, I got about a quart of air. What I took first was considerably nitrous, two measures of common air and one of this, occupying the space of $2\frac{1}{2}$ measures. The second pint was dephlogisticated; so that two measures of it, and five of nitrous air, occupied the space of two measures. The last was less dephlogisticated, being about one-half better than common air. At this time the air was generated with prodigious rapidity; the glass-tube through which it was transmitted was exceedingly red; and when, in changing the vessels, some of the vapour escaped into the air, it had the reddest appearance of any thing that I had ever seen of the kind.

Having saturated half an ounce of exceedingly good *quick-lime* with diluted spirit of nitre, dried it, and put it into a gun-barrel, I got from it about a pint and half of air, the first part of which was so far dephlogisticated, that it required an equal measure of nitrous air to saturate it. The second was no better than common air, and the third was equal to the first. In this process the air was produced very irregularly, sometimes coming in great quantities, and at other times the water would rush back into the tube.

I repeated the experiment on quick-lime, in a glass-phial and tube, when the whole quantity was so pure, that it required twice its bulk of nitrous air to saturate it. The produce of air, in this experiment, was as irregular as in the preceding. I could have wished to have treated the stone from which that lime was made in the same manner, but I had no opportunity.

From *lime fallen in air*, moistened with spirit of nitre, and treated as above, in a gun-barrel, I got near a pint of air, the greatest part of which came very rapidly, the fire being urged very much; and it was so far dephlogisticated, that two measures of it required five measures of nitrous air to saturate it. The last produce came very slowly, and it was so far nitrous, that two measures of common air, and one of it, occupied the space of less than two measures; that is, it was very nearly perfectly nitrous.

I also moistened with spirit of nitre a quantity of *lime that had been plunged in water*, in order to make lime-water, and got air from it in a gun-barrel, very irregularly, as before: one part of this air, which came almost at once, was dephlogisticated, so that two measures of this, and five of nitrous air occupied the space of $2\frac{1}{2}$ measures.

From two ounce-measures of pounded *marble*, treated as above, in a gun-barrel, I got about three quarts of air; but a very great proportion of it was fixed air, especially the last produce, which indeed was very little else; but towards the beginning of the process, the residuum was a little better than common air.

Repeating this experiment on marble, in a glass-phial with a ground-stopple and tube, I extracted from an ounce-measure of it about two pints of air, the greatest part of which was so highly dephlogisticated, that it took nearly three equal quantities of nitrous air to saturate it. Even the last produce was hardly to be distinguished from the first. What remained in the phial after the experiment swelled and broke it.

From *magnesia*, both calcined and uncalcined, I got, in a gun-barrel, a considerable quantity of air. From the calcined magnesia it was not much better than common air ; from the uncalcined it was more than twice as good. But very probably this difference may not be invariable.

I think it very probable that dephlogisticated air may be procured from any kind of earth with which the spirit of nitre will unite ; especially if it will likewise admit of a combination with fixed air or alkaline air, so that the spirit of nitre must expel the fixed air or alkaline air, before it can incorporate with it. Of substances of these kinds, besides those above mentioned, I tried *salt of tartar* and *wood-ashes*.

From half an ounce of *salt of tartar*, moistened with smoking spirit of nitre, and dried, I got, in a gun-barrel, about half a pint of air, the greatest part of which was fixed air, with a residuum so far dephlogisticated, as to be about three times as good as common air. The produce of air, in this experiment, was not very rapid, and it continued a long time. More would have been collected ; but that part of it escaped at the luting.

I moistened about half an ounce-measure of *ashes of wood*, carefully burned, first in an iron-ladle, and then in a crucible, with strong smoking spirit of nitre ; and, in a gun-barrel, I got from it about three pints of air, part of which was fixed, precipitating lime in lime-water, &c. and the rest was so pure, that it absorbed nearly three times its bulk of nitrous air. The last produce was very slow, and only about twice as good as common air.

From an ounce-measure of ashes of *pit-coal*, burned with all possible care, and treated in the same manner as above, I got about three quarts of air, one-third of which was fixed air, precipitating lime in lime-water; but the residuum was strongly nitrous, especially at the last. It is observable enough, from their dark colour (of which no burning, I believe, will divest them) that, in general, ashes of pit-coal contain much more phlogiston than ashes of wood.

N.B. In this, and other processes, it will be observed, that fixed air was procured from substances, which can hardly be thought to have contained it before: which affords a presumption, that it is not an acid *sui generis*, but a modification of the nitrous acid.

Clay is a substance altogether different from calcareous earth, and is not supposed, I believe, to contain any air at all. Of that species of it, which is called *tobacco-pipe-clay*, and which, I believe, is the purest of all, I got from Dr. Higgins, a quantity in powder; and moistening it with spirit of nitre, I observed that no more heat or effervescence was produced than the mixing of it with water would have occasioned.

Putting it, when dry, into a gun-barrel, I extracted from it, by a strong heat, about two ounce-measures of air, which being pretty readily absorbed by water, neither affecting common air, nor being affected by nitrous air, and extinguishing a candle, I concluded to be fixed air. Repeating the experiment, I got the same produce, only observing, that the fixed air made lime-water turbid, the most certain test, I believe, of the presence of fixed air; and the last produce was highly nitrous. Imagining that this produce might have come from the phlogiston of the iron, I resolved to repeat the experiment once more, with all possible care, in a glass-phial, with a ground-stopple and tube.

I did so, and took the produce at eight different times. The first and second quantities had a good deal of fixed air in them; the residuum of the first was a little diminished by nitrous air, almost as much as air in which a candle had burned out, which might be owing, in part, to the common air contained in the phial. The residuum of the second quantity, on the other hand, diminished common air a little; so that two measures of common air, and one of this, occupied the space of $2\frac{1}{3}$ measures. Of the third quantity, two measures required three measures of nitrous air to saturate it; so that it was pretty highly dephlogisticated. Of the fourth, two measures, and three of nitrous air, occupied the space of $1\frac{3}{4}$ measures. The fifth was of the same quality with the third. The sixth required twice its quantity of nitrous air to saturate it. The seventh was not quite so pure as the sixth; and the eighth neither affected common air, nor was affected by nitrous air, being what I term *phlogisticated air*. As some part of this produce was nitrous air, it is evident that the phlogiston necessary to constitute it must have been in the clay, and not in the vessel containing it, which was of glass.

Having by me a quantity of *Stourbridge clay*, I had the curiosity to repeat the experiments with this species, to see whether there would be any material difference in the result. Using the gun-barrel, I received the air in four portions. The first was fixed air, making lime-water turbid, and being more than half-absorbed by water; the second was about as good as common air, and the fourth was considerably nitrous.

To avoid the effect of the gun-barrel, I then put the clay into a phial, with a glass-stopple and tube, and putting it into a sand-heat, I received the air, for the sake of greater exactness, in ten different portions, about one-half of an ounce-measure each. The first produce was half-absorbed by water, with a residuum so far nitrous, that two measures of common air and one of this, occupied the space of $2\frac{1}{2}$ measures. The second and third portions were almost wholly fixed air, precipitating lime in lime-water, and not at all affecting common air, or being affected by nitrous air. Of the fourth I have no account. The fifth was so far dephlogisticated, that two measures of it, and three of nitrous air, occupied the space of $2\frac{1}{2}$ measures. The sixth and seventh produce were, as nearly as possible, common air. The ninth was so far nitrous, that two measures of common air, and one of this, occupied the space of $2\frac{1}{2}$ measures; and the tenth diminished common air still less.

It is evident, from the course of this process, that phlogiston must have been contained in the clay, and have been disengaged at different times, according as the heat affected different parts of the mixture. Had the whole of this produce been taken together, it would have been about the standard of common air mixed with fixed air; which shews the importance of taking the produce in different vessels, and examining them separately; a practice which the reader will find I have often had recourse to with great advantage.

As a great degree of heat cannot be applied to any thing contained in a glass-vessel, without melting it, and I was willing to know what would be the effect of more heat on this very clay when the above mentioned experiment was over, I took it out of the phial, and put it into a gun-barrel; when I got a considerable quantity of air from it. Part of it was fixed air, precipitating lime in lime-water, and the remainder was like the residuum of fixed air, or phlogisticated common air, extinguishing a candle, and being neither affected by nitrous air, nor affecting common air. I did the same thing with the *tobacco-pipe-clay*, which remained after the experiment above recited, and had nearly the same result. The first produce was of the same degree of purity with common air, and the next was a little affected by nitrous air.

From a quantity of *gypsum*, which I procured in the form of powder, I got a quantity of fixed air in a gun-barrel; and from the same, moistened with spirit of nitre, and treated in the same manner, I got a little fixed air, with a great proportion of nitrous air, almost as strong as any. But suspecting that this gypsum was not pure, I got of Dr. Higgins a piece of that kind of which the finest plaister is made, and from this, mixed with spirit of nitre, I got a considerable quantity of air, part of which was fixed air, and the remainder neither affected common air, nor was affected by nitrous air, and extinguished a candle. At last the air was nitrous, as I suspect, from the gun-barrel.

Being rather surprized that this kind of earth, which had the appearance of being very free from phlogiston, should yield air of no better quality than this, I repeated the experiment, by taking the produce of air at several times, as in former experiments, moistening the earth with a stronger spirit of nitre than before; and instead of a gun-barrel, made use of a phial with a ground-stopple and tube. The quantity of air produced in this manner, was about two ounce-measures, from an ounce-measure of the plaister, and I received the air in four different parts.

The first was a little diminished by nitrous air, being, I suppose, in a great measure, common air not quite expelled from the phial; and the second was strong nitrous air, perhaps from some phlogistic matter accidentally mixed with the ingredients. I am the more induced to think so, because the third and fourth produce was so highly dephlogisticated, that one measure of each took five measures of nitrous air to saturate it; so that they were each four times as good as common air.

After the preceding experiments, there remained only the *crystalline* and *talcky earths*, that are essentially different from each other; and each of these also yielded dephlogisticated air, when they were treated in the same manner with the earths above mentioned, but in a very small quantity.

When I took common *flints*, as they are dug out of the ground, part white and part black, moistening the powder of them with spirit of nitre, as before, and using a gun-barrel, I got fixed air, with a great proportion of nitrous air; that which came over the first being like the residuum of fixed air, extinguishing a candle, but being not readily absorbed by water.

After this I got some *flints carefully calcined in close vessels*, by Dr. Higgins, and having pounded a quantity of them, and moistened the powder with spirit of nitre, I put it into a glass-phial with a ground stopple and tube; and applying, at first, the flame of a candle only, the air I got was in a very small quantity; but it precipitated lime in lime-water, and diminished common air a little.

I then put the same apparatus into a sand-heat, when I got, in all, as much air as twice the bulk of the materials. Part of it precipitated lime in lime-water, but the rest of the produce was considerably better than common air; and the last was so good, that it took two measures of nitrous air to saturate it.

Lest this air might come from some extraneous matter, mixed with the powder of flint, I put some fresh spirit of nitre upon the same materials, without taking them out of the phial, after I had found that they would yield no more air from the first process, and I replaced the phial in the same sand-heat. The air first produced in this second process was but little diminished by nitrous air, but the rest was almost as pure as any that I had ever got before. The quantity of it, however, was not more than the bulk of the materials.

N.B. When, in this experiment, the bubbles of air burst, after getting through the water, a whitish cloud issued from them, as in the rapid production of nitrous air, and as in the produce of dephlogisticated air from the flowers of zinc above mentioned, but in a much less degree.

I repeated the same process not less than half a dozen times, putting fresh spirit of nitre upon the same materials, without taking them out of the phial, but the result was always the same; the first produce of air being always phlogisticated, then (after an interval in which nothing but the pure vapour of spirit of nitre came over) the remainder being the dephlogisticated air above mentioned.

To complete this course of experiments, I, in the last place, put strong spirit of nitre into a phial, filled with transparent *Muscovy talc*, such as opticians make use of for confining microscopic objects. In this process every thing went on in the very same manner as with the calcined flint; the first produce being phlogisticated air, or air of such quality as neither to affect common air, nor be affected by nitrous air, then the pure vapour of the spirit of nitre; and lastly, about an ounce-measure of air, about five times as good as common air. The pieces of talc which had been contiguous to the sides of the phial appeared to be a little whitened after the experiment, but the rest looked as if they had never been used in that manner; being as transparent as before, and of as firm a texture, but seemingly more flexible; so that those pieces, when handled all together, felt like soft feathers.

It is sufficiently evident from these experiments, that dephlogisticated air is produced from all kinds of earth mixed with spirit of nitre, only that a greater quantity of air is produced from some than from others; the advantage in this respect being on the side of the metallic and calcareous earths.

I would observe, that this process seems to furnish a pretty accurate test, perhaps the most accurate hitherto known, of the presence of phlogiston in bodies. Perhaps no species of air can be produced without a certain portion of phlogiston; but probably the nitrous acid itself always contains sufficient for the purpose of dephlogisticated air. But nitrous air contains so much phlogiston, that I think it cannot be produced unless the materials themselves contain it in a very considerable degree. Thus I have no doubt but that *white arsenic*, though it may be thought to contain no phlogiston, really does contain a considerable quantity of it; whereas, if the air be highly dephlogisticated, I think it may be considered as the most satisfactory proof we are yet acquainted with, that the substance contains no phlogiston at all.

I shall close this section with an account of the extraction of pure air from other substances besides *mercurius calcinatus*, and *red lead*, without the addition of spirit of nitre. Of this kind I have only found two substances, viz. *sedative* salt, and *Roman vitriol slightly calcined*, which I had from Dr. Higgins; besides common *salt-petre*, which is known to contain the acid of nitre in itself. From the two former I extracted air by means of a burning lens in quicksilver.

The *sedative salt* is not very manageable in this process; but, with some difficulty, I did extract from it a small quantity of air, in which a candle burned as in common air, and which was diminished as much as common air by nitrous air. At another time, the air which I extracted from this substance was not diminished by nitrous air quite so much as common air is.

N.B. The quantity of air was always very small, not more than the bulk of the materials.

From the *Roman vitriol* I also got but a small quantity of air. The first that I got was diminished by nitrous air, exactly as much as common air. I repeated the experiment, and the air which I then got was diminished by nitrous air considerably more than common air. The result of these experiments rather surprizes me, as, after many trials made with a view to it, I could get no such air from any species of *factitious vitriol*, calcined or uncalcined. There must certainly have been some nitrous acid in that Roman vitriol.

The readers of my former publications on this subject will remember, that I was exceedingly puzzled with the experiments which I made to extract air from *salt-petre* in a gun-barrel; the results appearing to me very extraordinary, and well worth attending to, as they might lead to considerable discoveries. (Vol. I. p. 155.) In fact, there was sufficient reason for the conjecture; but the method which I then took to extract air from this substance was ill adapted to make it yield its genuine produce. I had not, however, at that time, thought of any other.

The air I first got admitted a candle to burn in it with a very strong flame, and with a crackling noise. Also, though, after having stood a whole year in water, it became quite noxious, yet by agitation in fresh water, it was perfectly restored; so that a candle would burn in it again. At the time of my last publication, I conjectured that this air was *phlogisticated nitrous air*; but now I think it must have been dephlogisticated air, though produced in a gun-barrel, in which the spirit of nitre, by dissolving the iron, would be very apt to deprave the air; and accordingly, in repeating this experiment some time afterwards, I got air that extinguished a candle.

I was much puzzled, at that time, to account for the very different results of what was, to appearance, the same experiment; but I do not wonder at it now. I imagine that, in the former case, the air was produced very rapidly, and therefore that there was not time for the spirit of nitre to act upon the iron; and consequently the salt-petre gave its natural produce : whereas, in the latter case, a mixture of nitrous air (produced by the solution of iron in the nitrous acid, disengaged from the salt-petre) had thoroughly depraved the air. I advance this with the more certainty, as I have found that salt-petre, heated in a glass-vessel, yields very pure dephlogisticated air; its own earth, and the spirit of nitre which it contains, being capable, by heat, of forming that kind of union of those two principles which the constitution of that air requires ; and this, I think, is a pretty remarkable circumstance.

It may be worth while to observe, that I began my experiments upon nitre in quicksilver; but that the air produced in this manner was nitrous, occasioned by the solution of the quicksilver, as in the former case, by the solution of the iron in the spirit of nitre disengaged in the operation. A copious white fume issued from the nitre in the course of this experiment, like that which attends the rapid production of nitrous air from metals.

When I had recourse to my tall glass-vessels (fig. *d.*) I used an ounce of salt-petre pounded; and filling the vessel up to the mouth with pounded flint, I took the produce of air at nine times, each about three quarters of an ounce-measure. The first produce was not quite so good as common air, the second was of the same degree of purity with common air, the third rather worse; but the fourth was so far dephlogisticated, that one measure of it, and two of nitrous air, occupied the space of one-fifth less than one measure. The fifth produce was still better; for one measure of it, and two of nitrous air, occupied the space of half a measure. The ninth was about the same degree of purity; and the rest, I presume, were not much different.

Being desirous of knowing what kind of air was produced by the explosion of gun-powder, I, for that purpose, mixed equal quantities of brimstone and salt-petre, both finely pounded, and put them into a tall glass-vessel. The production of air was very rapid and copious, and so highly nitrous, that two measures of common air, and one of this, occupied the space of $2\frac{1}{4}$ measures. Since the produce of air from spirit of nitre and charcoal is the very same with this, viz. nitrous air, it cannot be doubted but that nitrous air is also produced in the explosion of gun-powder, which is composed of those ingredients; the spirit of nitre not being destroyed, or so far decomposed as that its acid nature is lost, but only entering into the composition of this species of air.

SECTION V.

MISCELLANEOUS OBSERVATIONS ON THE PROPERTIES OF DEPHLOGISTICATED AIR.

I ENDEAVOURED, in a variety of ways, to find the specific gravity of dephlogisticated air, by carefully weighing the materials before and after the production; and though this is by no means an exact method of ascertaining this circumstance, and I had recourse to better methods afterwards, the experiments may be worth reciting.

Having put into a gun-barrel two ounces four pennyweights of red lead, I extracted from it twenty ounce-measures of dephlogisticated air, receiving it in water; and the residuum, collected with all the care that I could apply, weighed 1 oz. 16 dwt. 18 gr.; so that twenty ounces of air ought to have weighed 7 dwt. 6 gr. which is beyond all proportion; so that this method must be very uncertain: besides, no allowance was made, nor could well be made, for the fixed air which the red lead yielded, and which is the heaviest species of air that we are yet acquainted with. At other times I have found that red-lead was changed into a real lead, when I was attempting the same thing in this way.

A second attempt came a little nearer the truth. I weighed an ounce of red lead, moistened it with smoking spirit of nitre, and dried it, when it weighed 1 oz. 6 dwt. 12 gr. I then divided the whole quantity into two equal parts, and put one of them into a gun-barrel, in order to collect the air, and the other I put into a crucible, to be exposed to the same degree of heat. The former yielded twenty-two ounce-measures of air, after the fixed air was pretty well washed out of it. It was about five times as good as common air. The latter had lost nineteen grains in weight, being just so much less than half an ounce; so that the twenty-two ounce-measures of air should have weighed nineteen grains, which is certainly a great deal too much: besides, in this experiment, as in the former, no account could be taken of the fixed air.

Finding these methods to fail, I had recourse to that which was used by Mr. Cavendish in weighing fixed and inflammable air, and which is more accurate than the method which I had used before (viz. filling a Florence-flask with the different kinds of air, and weighing them in it) because, as the flask must be first filled with water, one cannot be sure, though every possible precaution be taken, that the water has been equally drained from it after each experiment: otherwise there would be a considerable advantage in this method; because the quantity of air may be accurately known. But though this cannot be done with precision in a bladder, as used by Mr. Cavendish, because the degree of distension cannot be measured with much accuracy, yet this circumstance is more than counterbalanced by being able to change the air with compressing the bladder, without wetting it.

I therefore took a glass-tube about nine inches long, and fastening it to the neck of a bladder, which, with such a degree of distension as I could give it, in the manner in which the experiment was made, contained fifty-five ounce-measures, or one pennyweight nine grains of common air. The tube was so fastened, that I could take it out at pleasure; and having the bladder thus prepared, I carefully compressed it, then filling it in part with that kind of air which I was about to weigh, I compressed it again, and then filled it intirely; so that I was

pretty confident that the air within the bladder contained
very little common, or any other kind of air. In this
manner I proceeded to weigh *dephlogisticated air*, and at
the same time *nitrous air*, and *air phlogisticated with
iron filings and brimstone*, which I take for granted is the
same thing with air phlogisticated by any other process.

The following short table exhibits the result of all
these experiments at one view.

The bladder, filled with					dwts.	gr.
phlogisticated air, weighed -	-	-	-	-	7	15
——————— nitrous air	-	-	-	-	7	16
——————— common air	-	-	-	-	7	17
——————— dephlogisticated air -	-	-	-	7	19

The result agrees sufficiently well with my former
observations, though they were not made with so much
accuracy, viz. that both nitrous air, and air diminished
by phlogistic processes, are rather lighter than common
air; and it is consonant to this, that, in the present
experiment, dephlogisticated air appears to be a little
heavier than common air.

Comparing these observations with that of the extreme
lightness of inflammable air, ascertained by Mr. Cavendish,
it should seem that the less phlogiston any kind of air
contains the heavier it is, and the more phlogiston it
contains the lighter it is; though this is by no means
the case with solid substances, and indeed it is rather
unfavourable to this hypothesis, that nitrous air should
not be lighter than dephlogisticated air; for it should
seem, by its property of phlogisticating common air, that
it should itself contain a greater proportion of phlogiston.
Also, in the above mentioned processes for making air,
the more phlogiston there is in the substances moistened
with spirit of nitre, the more certain it is that the produce
will be nitrous air; as the less phlogiston they contain,
the more certain it is that the produce will be pure air.

But I suspect that there is a farther difference in the *mode* in which phlogiston is combined with spirit of nitre, in the constitution of nitrous air.

In this experiment, the dephlogisticated air was so pure, that one measure of it, and two of nitrous air, occupied the space of $\frac{4}{5}$ of a measure. Had the air been more pure, it would, no doubt, have been specifically heavier still.

It should be observed, that sufficient time ought to be allowed to get dephlogisticated air intirely free from fixed air before it is weighed; and as this requires time, and perhaps may never be done completely, it may be suspected that the additional weight of this kind of air is owing to a mixture of fixed air. But common air also contains a great proportion of fixed air, and the dephlogisticated air, on which I made this experiment, had been produced, at least the far greatest part of it, and had been exposed to water, some weeks. It is, however, sufficiently evident, that dephlogisticated air doth become better by standing in water; owing, probably, to its depositing more fixed air in those circumstances.

Having at one time made a large reservoir of dephlogisticated air, for the purpose of experiments, I found that, in about ten days, from being $4\frac{1}{2}$, it had become $5\frac{1}{2}$ better than common air. Standing in pure water must be a surer method of getting the purest dephlogisticated air, than *agitation* in water; for, though the latter method will enable the water to absorb the fixed air faster, and therefore a little agitation at the first will be very useful, in order to expedite the purification of it; yet, as I have found (vol. I. p. 158) that agitation in the purest water will, in time, injure common air; the same operation may be supposed to injure dephlogisticated air also; and indeed I have already observed, that having agitated in water a quantity of dephlogisticated air, a candle burned in it only as in common air, and not with that vivid flame with which it burns in this air when it is purer.

I have not made many experiments on the mixture of dephlogisticated air with the other kinds of air, because the analogy which it bears to common air is so great, that I think any person may know before-hand, what the result of such experiments would be. It is pleasing, however, to observe how readily and perfectly dephlogisticated air mixes with phlogisticated air, or air injured by respiration, putrefaction, &c. each tempering the other; so that the purity of the mixture may be accurately known from the quantity and quality of the two kinds of air before mixture. Thus, if one measure of perfectly noxious air be put to one measure of air that is exactly twice as good as common air, the mixture will be precisely of the standard of common air.

I observed also, in making this experiment, that after mixing one measure of each of these kinds of air, they made exactly two measures; so that there was neither any increase nor diminution of quantity in consequence of the mixture, as is the effect of mixing nitrous air with either common or dephlogisticated air.

It may hence be inferred, that a quantity of very pure air would agreeably qualify the noxious air of a room in which much company should be confined, and which should be so situated, that it could not be conveniently ventilated; so that from being offensive and unwholesome, it would almost instantly become sweet and wholesome. This air might be brought into the room in casks; or a laboratory might be constructed for generating the air, and throwing it into the room as fast as it should be produced. This pure air would be sufficiently cheap for the purpose of many assemblies, and a very little ingenuity would be sufficient to reduce the scheme into practice.

I easily conjectured, that inflammable air would explode with more violence, and a louder report, by the help of dephlogisticated than of common air; but the effect far exceeded my expectations, and it has never failed to surprize every person before whom I have made the experiment.

Inflammable air requires about two-thirds of common air to make it explode to the greatest advantage; and if a phial, containing about an ounce-measure and half, be used for the experiment, the explosion with common air will be so small, as not to be heard farther than, perhaps, fifty or sixty yards; but with little more than one-third of highly dephlogisticated air, and the rest inflammable air, in the same phial, the report will be almost as loud as that of a small pistol; being, to judge by the ear, not less than forty or fifty times as loud as with common air.

The orifice of the phial in which this experiment is made, should not much exceed a quarter of an inch, and the phial should be a very strong one; otherwise it will certainly burst with the explosion. The repercussion is very considerable; and the heat produced by the explosion very sensible to the hand that holds it. I have sometimes amused myself with carrying in my pocket, phials thus charged with a mixture of dephlogisticated and inflammable air, confined either with common corks or ground-stopples, and I have perceived no difference in the explosion, after keeping them a long time, and carrying them to any distance.

The dipping of a lighted candle into a jar filled with dephlogisticated air is alone a very beautiful experiment. The strength and vivacity of the flame is striking, and the heat produced by the flame, in these circumstances is also remarkably great. But this experiment is more pleasing, when the air is only little more than twice as good as common air; for when it is highly dephlogisticated, the candle burns with a crackling noise, as if it was full of some combustible matter.

It may be inferred, from the very great explosions made in dephlogisticated air, that, were it possible to fire gunpowder in it, less than a tenth part of the charge, in all cases, would suffice; the force of an explosion in this kind of air, far exceeding what might have been expected from the purity of it, as shewn in other kinds of trial. But I do not see how it is possible to make this application of it. I should not, however, think it difficult to confine gunpowder in bladders, with the interstices of the grains filled with this, instead of common air; and such bladders of gunpowder might, perhaps, be used in mines, or for blowing up rocks, in digging for metals, &c.

Nothing, however, would be easier than to augment the force of fire to a prodigious degree, by blowing it with dephlogisticated air instead of common air. This I have tried, in the presence of my friend Mr. Magellan, by filling a bladder with it, and puffing it, through a small glass-tube, upon a piece of lighted wood: but it would be very easy to supply a pair of bellows with it from a large reservoir.

Possibly much greater things might be effected by chymists, in a variety of respects, with the prodigious heat which this air may be the means of affording them. I had no sooner mentioned the discovery of this kind of air to my friend Mr. Michell, than this use of it occurred to him. He observed that possibly *platina* might be melted by means of it.

From the greater strength and vivacity of the flame of a candle, in this pure air, it may be conjectured, that it might be peculiarly salutary to the lungs in certain morbid cases, when the common air would not be sufficient to carry off the phlogistic putrid effluvium fast enough. But, perhaps, we may also infer from these experiments, that though pure dephlogisticated air might be very useful as a *medicine*, it might not be so proper for us in the usual healthy state of the body: for, as a candle burns out much faster in dephlogisticated than in common air, so we might, as may be said, *live out too fast*, and the animal powers be too soon exhausted in this pure kind of air. A moralist, at least, may say, that the air which nature has provided for us is as good as we deserve.

My reader will not wonder, that, after having ascertained the superior goodness of dephlogisticated air by mice living in it, and the other tests above mentioned, I should have the curiosity to taste it myself. I have gratified that curiosity, by breathing it, drawing it through a glass-syphon, and, by this means, I reduced a large jar full of it to the standard of common air. The feeling of it to my lungs was not sensibly different from that of common air; but I fancied that my breast felt peculiarly light and easy for some time afterwards. Who can tell but that, in time, this pure air may become a fashionable article in luxury. Hitherto only two mice and myself have had the privilege of breathing it.

Whether the air of the atmosphere was, in remote times, or will be in future time, better or worse than it is at present, is a curious speculation; but I have no theory to enable me to throw any light upon it. Philosophers, in future time, may easily determine, by comparing their observations with mine, whether the air in general preserves the very same degree of purity, or whether it becomes more or less fit for respiration in a course of time; and also, whether the changes to which it may be subject are *equable*, or otherwise; and by this means may acquire *data*, by which to judge both of the past and future state of the atmosphere. But no observations of this kind having been made, in former times, all that any person could now advance on this subject would be little more than random conjecture. If we might be allowed to form any judgment from the length of human life in different ages, which seems to be the only *datum* that is left us for this purpose, we may conclude that, in general, the air of the atmosphere has, for many ages, preserved the same degree of purity. This *datum*, however, is by no means sufficient for an accurate solution of the problem.

ALEMBIC CLUB REPRINTS.

Crown Octavo. Cloth. Uniform.

VOLUMES ALREADY PUBLISHED.

No. 1.—EXPERIMENTS UPON MAGNESIA ALBA, Quick-Lime and Other Alcaline Substances. By JOSEPH BLACK, M.D. 1755. 47 pp. Price 1s. 6d. net.

No. 2.—FOUNDATIONS OF THE ATOMIC THEORY: Comprising Papers and Extracts by JOHN DALTON, WILLIAM HYDE WOLLASTON, M.D., and THOMAS THOMSON, M.D. 1802-1808. 48 pp. Price 1s. 6d. net.

No. 3.—EXPERIMENTS ON AIR. Papers published in the Philosophical Transactions. By the Hon. HENRY CAVENDISH, F.R.S. 1784-1785. 52 pp. Price 1s. 6d. net.

No. 4.—FOUNDATIONS OF THE MOLECULAR THEORY: Comprising Papers and Extracts by JOHN DALTON, JOSEPH LOUIS GAY-LUSSAC, and AMEDEO AVOGADRO. 1808-1811. 52 pp. Price 1s. 6d. net.

No. 5.—EXTRACTS FROM MICROGRAPHIA. By R. HOOKE, F.R.S. 1665. 52 pp. Price 1s. 6d. net.

No. 6.—ON THE DECOMPOSITION OF THE ALKALIES AND ALKALINE EARTHS. Papers published in the Philosophical Transactions. By HUMPHRY DAVY, Sec. R.S. 1807-1808. 52 pp. Price 1s. 6d. net.

No. 7.—THE DISCOVERY OF OXYGEN. Part I. Experiments by JOSEPH PRIESTLEY, LL.D. 1775. 56 pp. Price 1s. 6d. net.

No. 8.—THE DISCOVERY OF OXYGEN. Part II. Experiments by CARL WILHELM SCHEELE. 1777. 46 pp. Price 1s. 6d. net.

Postage of any of the above to any part of the World, 2d. each extra.

IN THE PRESS.

No. 9.—ON THE ELEMENTARY NATURE OF CHLORINE. Papers published in the Philosophical Transactions. By HUMPHRY DAVY, Sec. R.S. 1810-1818.

WILLIAM F. CLAY, Publisher,
18 TEVIOT PLACE, EDINBURGH.

www.ingramcontent.com/pod-product-compliance
Lightning Source LLC
Chambersburg PA
CBHW031930060726
47496CB00008BA/2794